Dear Parent:
Your child's love of reading starts here!

Every child learns to read in a different way and at his or her own speed. Some go back and forth between reading levels and read favorite books again and again. Others read through each level in order. You can help your young reader improve and become more confident by encouraging his or her own interests and abilities. From books your child reads with you to the first books he or she reads alone, there are I Can Read Books for every stage of reading:

SHARED READING
Basic language, word repetition, and whimsical illustrations, ideal for sharing with your emergent reader

BEGINNING READING
Short sentences, familiar words, and simple concepts for children eager to read on their own

READING WITH HELP
Engaging stories, longer sentences, and language play for developing readers

READING ALONE
Complex plots, challenging vocabulary, and high-interest topics for the independent reader

I Can Read Books have introduced children to the joy of reading since 1957. Featuring award-winning authors and illustrators and a fabulous cast of beloved characters, I Can Read Books set the standard for beginning readers.

A lifetime of discovery begins with the magical words **"I Can Read!"**

*Visit www.icanread.com for information
on enriching your child's reading experience.*

Whether you love running wild or staying close to home,
this book is for you!
—J.B.

The National Wildlife Federation & Ranger Rick contributors: Children's
Publication Staff, Licensing Staff, and in-house naturalist David Mizejewski

Ranger Rick: I Wish I Was a Llama
Copyright © 2020 National Wildlife Federation. All rights reserved.
Manufactured in China. No part of this book may be used or reproduced in any manner whatsoever without written permission except in the case of brief quotations embodied in critical articles and reviews. For information address HarperCollins Children's Books, a division of HarperCollins Publishers, 195 Broadway, New York, NY 10007.
www.icanread.com
www.RangerRick.com

Library of Congress Control Number: 2019944433
ISBN 978-0-06-243229-2 (trade bdg.) —ISBN 978-0-06-243228-5 (pbk.)

Book design by Brenda E. Angelilli
21 22 23 SCP 10 9 8 7 6 5 4 ❖ First Edition

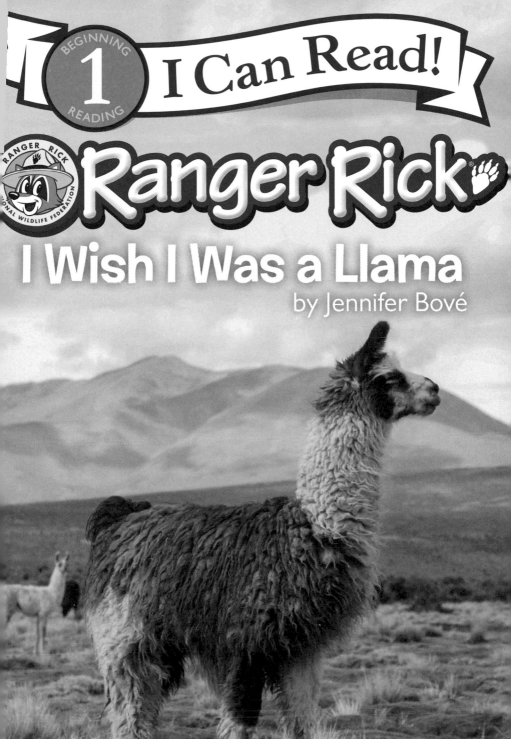

Ranger Rick

I Wish I Was a Llama
by Jennifer Bové

HARPER

An Imprint of HarperCollinsPublishers

What if you wished you were a llama?

That could be fun!

But llamas live with people.

Maybe you'd rather run WILD!

You might like to be the llama's

wild cousin, a guanaco (wah-nah-ko).

How are llamas and guanacos alike?

How are they different? Find out!

Where would you live?

Wild guanacos live far from people
in the Andes Mountains of South America.
Guanacos roam freely
around rocky, steep mountain slopes.
The weather there is cool all year.

Llamas are domesticated.

That means they live with people.

They live on farms
all over the world.

Llamas are not strong enough
to carry people the way horses do.
But they can carry backpacks
for people on hiking trips.

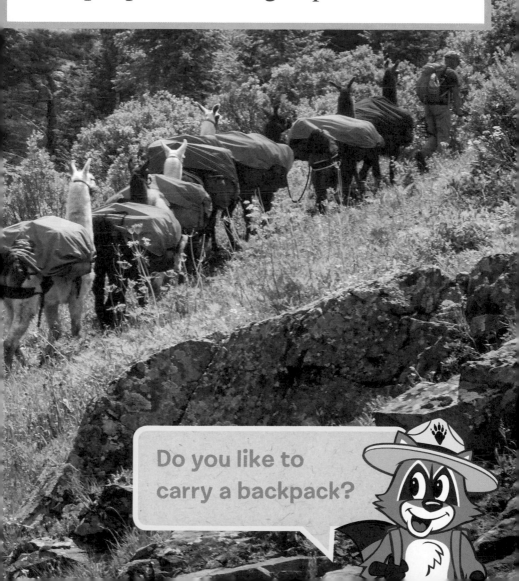

Do you like to
carry a backpack?

What would your family be like?

Llama and guanaco families

are called herds.

About twenty animals live in a herd.

There is usually one dad
with many moms and babies.
A baby llama or guanaco
is called a cria (cree-uh).

How many people are
in your family?

How would you learn to be a llama?

A cria learns by watching its herd.

In the wild, guanacos must be brave.

They chase hungry foxes away.

With practice, a cria can, too.

On farms, llama crias see their herd working happily with humans. They learn that people are friendly.

What would you eat?

Llamas and guanacos eat tough grass, shrubs, and woody plants.

On farms, people feed llamas hay.

Tough plants need a lot of chewing.

These animals actually throw up food so that they can chew it again!

This is called chewing cud.

Llamas and guanacos do not need
to drink much water.

They get most of the water they need
from the plants they eat.

These animals have another cousin that drinks very little: the camel!

How would you wash up?

Llamas and guanacos roll in dust.
Rolling keeps itchy bugs away
and fluffs up their coats.
Fluffy hair feels cool in the summer
and warm in the winter.

How would you talk?

Llamas like being together, and so do guanacos.

But they do not talk much.

Do you hum when you feel happy?

Mostly, they are calm
and quiet animals.
Sometimes they click or gurgle,
and they hum to say, "I'm happy."

Llamas and guanacos are louder when they are unhappy. They groan or squeal to say, "I don't like that!"

Angry llamas lay their ears flat and spit at each other!

Where would you sleep?

Llamas and guanacos sleep
on the ground in open spaces.
They like being able to wake up
and see all around them.
They sleep with their heads up
and their legs folded under them.
This is called kushing.

Would you want to sleep with
your head sticking up?

Llamas and guanacos are a lot alike.

But they live in different ways.

Llamas are domesticated.

They get to hang out with people.

26

Guanacos live far from humans.

They run wild their whole lives.

What if you were a llama

or a guanaco?

Either could be cool for a while.

But do you want to chew cud?

Roll in dust?

Spit when you're mad?

Luckily, you don't have to.

You're not a llama or a guanaco.

You're YOU!

Did You Know?

🐾 Llamas and guanacos have another wild relative in South America called the vicuña (vik-oon-ya). It looks like a guanaco, but smaller. Domesticated vicuñas are called alpacas.

🐾 Llamas, guanacos, vicuñas, and alpacas are all related to camels.

🐾 A llama can spit a distance of over fifteen feet (about five meters) and hit its target (usually another llama).

Fun Zone

Llamas and their relatives look a lot alike.
Use the clues below to guess which animal is shown in each picture. The answers are upside down at the bottom of the page.

🐾 Camels are bigger than llamas. They have short ears and humped backs.

🐾 Llamas have long, shaggy hair. Their ears are long and curved.

🐾 Guanacos have shorter hair than llamas. They have long legs and long ears.

🐾 Alpacas are smaller than llamas. They have woolly hair and short, pointy ears.

🐾 Vicuñas look like guanacos, but they are smaller and have very skinny necks.

Answers: 1. Guanaco 2. Camel 3. Llama 4. Alpaca 5. Vicuña

31

Wild Words

Cria: a young llama

Cud: food that is thrown up and chewed again

Domesticated animal: an animal that lives with and is cared for by people

Guanaco: a wild South American animal; related to the camel and llama

Herd: a family of guanacos or llamas

Kushing: the way llamas lie down with their heads up and legs folded

Llama: a domestic South American animal; related to the camel

Wild animal: an animal that lives in nature and is not cared for by humans

Dig Deeper

WANT TO FIND OUT EVEN MORE ABOUT LLAMAS?

Check out the Ranger Rick website: www.RangerRick.com
SEARCH: llama